# Build Rabbit Housing

## by Bob Bennett

Properly housed and fed domestic rabbits can provide you with delicious, high-quality meat with little expenditure of time, space, or money. Rabbits are small and quiet, and they can be kept essentially odorfree with the proper approach to their housing and care. You can store them on the hoof until you are ready to prepare them for the table.

The equipment you need to build and outfit a rabbitry is quite limited: hutches, feeders, waterers, and nest boxes. Essential hand tools to build the rabbitry are also few; most households have all but one or two. No power tools are required; nor do you need a large outlay of cash or time. You can construct and outfit your rabbitry as both allow. Interruptions to the construction process will present no problem — you can drop the work and resume it at any point. You can build indoors or out; the materials make no dust or dirt and require no loud banging or sawing. Everything is quite clean and quiet. And simple. If you follow the instructions, your very first efforts will produce a product that functions as well and looks as good as any built by experienced professionals.

This bulletin explains what you need to build and outfit a small noncommercial rabbitry and how to do it. Should you find raising rabbits very satisfying and you want to expand your facilities, you simply add more of the same.

# The All-Wire Hutch

Successful rabbit raisers consider no other materials for a rabbit hutch than wire and metal. Among other activities, rabbits gnaw and urinate, so you can dismiss wood immediately. If you use wood, they will eat some of it and foul the rest of it. Wire and metal cost less than lumber; no expensive hinges or other hardware are required.

If you need only one or two hutches, perhaps for pet rabbits, buy them from your local farm supply store or send away to a supplier listed in this bulletin. That's because the welded wire fabric you need is expensive in small quantities (if it is available at all), and it will be cheaper for you to buy one or two than to build them.

To build your own, here are the tools and materials you will need.

## Tools

• One pair of heavy-duty, 7″ or 8″ wire-cutting pliers, preferably with flush-cutting jaws.

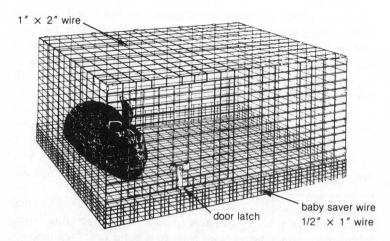

1″ × 2″ wire

door latch

baby saver wire
1/2″ × 1″ wire

*The basic all-wire hutch. Front, top, back, and sides are 1″ × 2″ wire. Floor is 1/2″ × 1″ wire. Door is positioned off center to allow a feeder and waterer at right.*

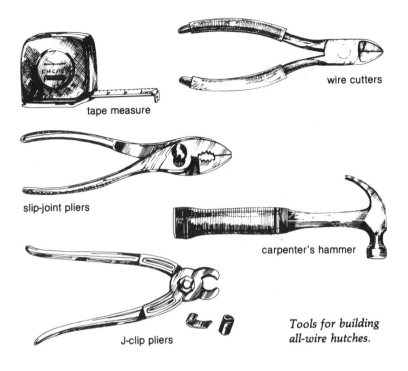

tape measure

wire cutters

slip-joint pliers

carpenter's hammer

J-clip pliers

*Tools for building
all-wire hutches.*

- One pair of ordinary slip-joint pliers.
- One tape measure, preferably retractable steel. A 12' tape is best but a shorter one will do.
- An ordinary carpenter's hammer.
- One pair of special J-clip pliers, available from your farm supply store or a supplier listed at the end of this bulletin.
- A short length of 2 × 4 lumber; about 3' long is fine.

## Materials

- A supply of 1" × 2" welded wire fencing, 14 gauge, sometimes called *turkey wire.* Dimensions will vary as described below.
- A supply of 1/2" × 1" welded wire, 14 or 16 gauge, with variable dimensions as described below.
- A supply of J-clips, available where you obtain the J-clip pliers.
- A door latch for each hutch.
- A door hanger for each hutch.

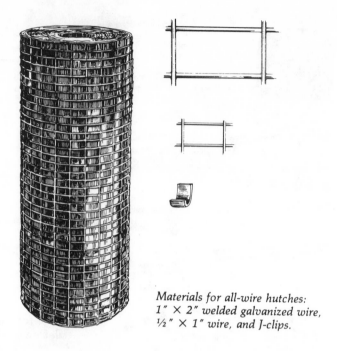

*Materials for all-wire hutches:*
*1″ × 2″ welded galvanized wire,*
*½″ × 1″ wire, and J-clips.*

Before we actually start hutch construction, let's discuss the tools and materials for a good understanding of what is involved.

## About the Tools

The wire-cutting pliers need to be large enough to give substantial leverage for cutting through 14-gauge wire. A good cutting tool, made of high-quality steel, isn't cheap. If you plan to make more than a few hutches, spend enough to get a good pair. Ideally, the jaws will be fairly narrow so they can get a good bite on the 1/2″ × 1″ wire. Some rabbit raisers buy a second, smaller pair for working with the smaller mesh wire if they can't find a larger pair with narrow jaws. Unbeveled cutting jaws will make a flush cut; the result will be a smoother hutch with few sharp edges. Less leverage is required with the smaller mesh wire. Plastic handle grips will make your pliers more comfortable, and wearing a leather glove will help prevent blisters.

J-clip pliers are made for use with J-clips. When squeezed tightly around two parallel wires, they produce a clamp unsurpassed for hutch making. It is true that you can use hog rings or C-rings and hog ring pliers; you can even get away with twisting wire stubs or other short pieces of wire around cage sections. These items and techniques will work, but they are time-consuming and usually will not produce as attractive and durable a hutch.

A metal bending brake is probably more efficient than a hammer and 2 × 4 for bending wire, but it is hardly worth the investment unless you build an extremely large rabbitry. If you have access to one, you will save a little time. Power wire-cutting shears are also on the market, but their cost is justified only if you plan to build a great many hutches.

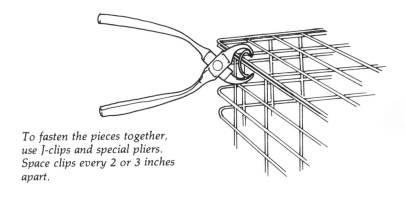

To fasten the pieces together, use J-clips and special pliers. Space clips every 2 or 3 inches apart.

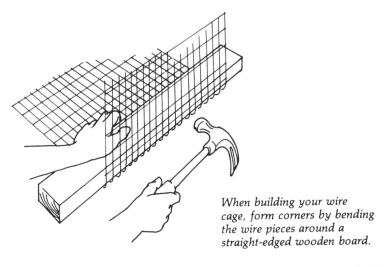

When building your wire cage, form corners by bending the wire pieces around a straight-edged wooden board.

## About the Wire

You will need welded wire fencing in both 14-gauge 1″ × 2″ and 1/2″ × 1″, or 16-gauge 1/2″ × 1″. The 14-gauge wire is heavier, better, and more expensive, but is more difficult to obtain in many areas. Instead of 1″ × 2″ wire you may substitute 1″ × 1″ or 1″ × 1/2″. Both of these will make a sturdier cage, but the cage will be overbuilt and not worth the extra expense and extra work (more cutting), unless you plan to raise the giant, extra-heavy breeds of rabbits, or you already have some of this wire available, or are offered some at a bargain price.

Welded wire comes in basically two types — galvanized *before* welding and galvanized *after* welding. Buy the latter. It costs more, but is more rigid and will last longer. Welded wire wears out by oxidation, and wire that is galvanized after welding has more material to oxidize. It will appear thicker, especially at the joints, than wire galvanized before welding, which is smoother to the touch. You can build with wire galvanized before welding but the results will not be as satisfactory.

Sometimes it is possible to find "aluminized" welded wire instead of galvanized. This is very good wire. If you can find some, it makes an excellent hutch.

Some rabbit breeders have experimented with vinyl-covered welded wire. Rabbits love to gnaw, however, and easily can strip the vinyl covering from the wire, which is ungalvanized and soon will rust.

Another point: there is both 1″ × 2″ wire and 2″ × 1″ wire. The former has wires every inch of the width of the roll. The latter has wires every 2 inches of the width of the roll. The 1″ × 2″ is better, but 2″ × 1″ will work. When calculating dimensions keep in mind that you lose an inch or two each time you cut this wire.

## Door Latches

As you can see from the picture, the recommended door latches are easy to fabricate from heavy galvanized metal that can be riveted or bolted together. These latches are so inexpensive, however, that buying them seems a better use of time, unless you happen to be adept at metal working and already have a supply of bolts or rivets and a riveter. The same can be said for the door hangers.

*Two types of galvanized metal door latches. Tabs at the top bend around door wire; clasp pivots to swing upward to left or right.*

## How Big to Make the Hutches

All-wire rabbit hutches consist of little more than boxes with 1" × 2" wire on front, back, top, and sides, and 1/2" × 1" wire on the bottom. How big to make the boxes and how many to make is up to you. Here are a few points for consideration.

**Single vs. Multiple Hutches.** Making single hutches at a time will give you total flexibility in hutch layout or rabbitry configuration. Rabbits tend to multiply, so your rabbitry is likely to grow and to change shape or design. Single hutches give you the flexibility you might want to get in and out of rabbit raising, to move, or to build totally new facilities. You can move single hutches easily. They appeal more to others who may purchase them from you later and use them in a different arrangement.

Multiple hutch units will save you time overall because you divide them with shared partitions and assemble them faster, with less cutting and fastening. They won't necessarily save you money on materials, however, because single shared partitions demand more costly 1/2" × 1" wire instead of 1" × 2". This additional expense can be debated, however, if you have purchased the 1/2" × 1" in 50' or 100' rolls, and have extra left over anyway. A good compromise is building hutches in 2-unit or 3-unit modules for openers, unless you are determined to start big and have firm plans for a great many hutches. You can also use leftover floor wire for nest boxes, as described later.

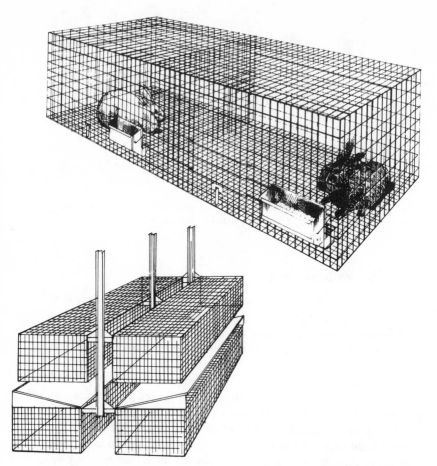

*Double-unit all-wire hutches. These hutches can be hung double-decker style, as shown.*

**Size Considerations.** While overall dimensions of hutches are up to you, as they can be made in just about any size up to maximum widths of available wire, there are three other considerations regarding size.

• How much room the rabbit needs. Provide nearly a square foot of floor space per pound of adult doe. A 2-1/2' × 3' hutch will accommodate a medium-size (meat breed) doe and her litter to weaning time. It will, of course, be sufficient for smaller breeds as well. Hutches for bucks and young growing stock can be half that size or a 2-compartment, 2-door hutch of the same size.

• How big a hutch is convenient for the rabbit keeper. For example, maximum depth front to back should not exceed 2-1/2'. If the hutch is deeper you won't be able to catch rabbits in the back unless you have the reach of a professional basketball player.

• Their location. If you are going to build hutches to fit inside a specific existing building, such as a garage or shed, you may have to adjust the hutch size accordingly (within the above guidelines).

## Buying Wire

For ease and speed of construction, the ideal situation is to purchase 3 sizes of welded wire. These include a roll of *floor wire* (1/2" × 1") of the desired width, a roll of *top wire* (1" × 2") of the same width as the floor wire, and a roll (or rolls) of *side* (and *front* and *rear*) wire of the desired height (usually 16" or 18" for the medium breeds, but 24" for the giants and as little as 12" or 15" for the dwarf and small breeds). One can even make a case for buying a fourth size of wire for doors. Available widths for welded wire are 12", 15", 18", 24", 30", 36", 48", 60", and 72".

# Two Approaches in Hutch Building

I'm going to describe 2 building schemes — the first if you wish to purchase materials for up to 10 hutches for breeding does of the medium breeds, and the second if you plan to build more than that of the same size. You will be able to decide from these two approaches which will be better for you. Because rolls of less than 100' will cost you more per foot, the first approach will save you money if you plan to build only 10 hutches.

## For up to 10 Hutches 2-1/2' × 3' × 16"

Purchase 100' of 1" × 2" wire, 36" wide. Purchase either 26' of 1/2" × 1" wire that is 36" wide or 31' of 1/2" × 1" wire that is 30" wide. If you plan to build all-wire nest boxes, as described later, you will need more of the 1/2" × 1" wire, so plan and buy accordingly. Buy 2 pounds of J-clips. There are about 450 to the pound. Each hutch takes about 90 clips, and you will bend a few out of shape as you get used to working with them.

Here's how to build these hutches one at a time.

**Step 1. Cut the Front, Back, and Top.** From the roll of 1" × 2" wire, cut a piece 62" long. Cut it flush and, while you're at it, cut the stubs off flush from the rest of the roll. Lay the piece on the floor so it curls down (humps up). Place your feet on one end and gently bend the other toward you, moving your feet forward as necessary to flatten the wire. Take care not to kink it; simply reverse the curve with enough pressure to take the bend out of the roll. Now turn it over before forming 90° corners. This is important so you don't bend against the welded joints, but *with* them. Most rolls of wire come with the 1" wires on top of the 2" wires as you view the roll before unrolling. If your wire is rolled the opposite way, then note it, and do not bend against the welds.

With the wire on the floor, measure 16" and lay your 2 × 4 board across it at that point. Stand on the 2 × 4 and pull the 16" section toward you gently. Hold the 16" end, reach down with

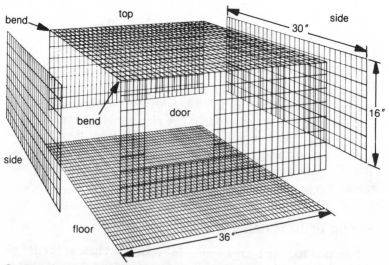

*Cutting plan for the individual all-wire hutch, 30" × 36" × 16". This hutch will house a medium-size meat-breed doe and her litter.*

your hammer, and gently strike each strand of wire against the 2 × 4 to make a 90° angle. Now turn around, measure 16" from the other end and repeat the bending process. You have just formed the front, the back, and the top. Set it aside.

**Step 2. Cut the Sides.** Next, cut 30" more off the roll and flatten the resulting 30" × 36" piece as above. Cut it flush and cut the stubs flush off the roll. (As you cut, notice that a slight flick of the wrist down and away from the welds will snap the wire off cleanly with little effort.) Measure 16" up (you are splitting the piece) and cut off flush. The remaining piece will be 30" × 18" (plus stubs). Measure 16" down and cut off 2" (plus stubs). The 2" "waste" strip will be from the center. You will use it later, so set it aside. You now have the sides (ends) of your hutch.

**Step 3. Cut the Floor.** If you bought 36" wide, 1/2" × 1" wire, cut off 30" flush and cut the stubs off the roll. If you bought 30" wire, cut off 36". Do *not* flatten this piece of floor wire, which curves or humps with the 1/2" wires up, unless it has an extreme curl (perhaps if it is cut from the inside of the roll, where it is curled tighter, in a smaller diameter). The idea is to have the floor of the hutch with the 1/2" wires up to provide a smoother surface for the rabbits, and to keep an upward spring that will eliminate sagging from the weight of the rabbits.

Now you are ready to assemble the hutch. You have cut out all the pieces except for the door and some more 2" strips that you will clip later to the 4 sides near the floor as a "baby saver" feature. More about that later.

**Step 4. Fasten the Sides in Place.** Place a J-clip in the jaws of the J-clip pliers and fasten the side sections (the 16" × 30" sections) to the front, top, and back sections. After a little practice you will find that squeezing the J-clips on requires a little flip of the wrist or a second squeeze to assure that the clip is tight. Use a J-clip every 4", starting with the corners. Make sure the vertical 1" wires are on the outside of the hutch. That way you will have horizontal wires on the inside where they will make neat, tight corners when fastened to the front-top-back section.

**Step 5. Fasten the Floor.** After fastening the ends to the front, back, and top, turn the hutch on its top and lay the floor wire on it with the curve and the 1/2" wires down (toward the top of the hutch). Remember that the 1/2" wires provide the smoother floor and that an upward spring will prevent floor sag.

Clip the floor wire on, starting in one corner, again using clips every 4". If the fit is too tight in a corner (this can happen if the 1" × 2" wire is made by a different manufacturer than the one who made the 1/2" × 1" wire), notch out the 1/2" corners of the floor wire.

From what looked like a flimsy beginning, you will find that you now have quite a sturdy hutch. Clipped together, the resulting wire box is extremely rigid.

Now you are ready to cut out the door opening and attach the door.

**Step 6. Cutting a Door Opening.** Door size and position are very important, so stop and ponder the situation. The door should be located to one side of the front (a 3' side) because you want to leave space on the front for attachment of a feeder and space to fill a waterer, be it a crock, bottle with tube or valve, or automatic watering fount or valve. The door opening must also be large enough to admit a nest box, which you will be putting in and taking out regularly.

If you use an all-wire nest box (which I recommend and explain how to build in this bulletin), a door opening that is 12" wide and 11" high is large enough yet not too large. Remember that, while the door must be conveniently large, you are cutting into the front of the hutch and weakening it somewhat, so a door that is un-necessarily large doesn't do the hutch a lot of good. In any event, be sure to consider the dimensions of the nest box and door together.

For this hutch, I recommend a door opening of 12" × 11" with a door that is 14" × 12" and swings up and in.

Stand the hutch on its back, with the front up and the floor closest to you as you approach it with your wire cutters. Measure 4" over from the left and 4" up from the bottom. Cut the bottom strands to the right to make a 12" opening. Cut up 11" on each side and 12" across the top. *Very important:* Do not cut these strands flush but leave stubs about 1/2" long all the way around the opening. Once you have the opening cut out, use your slip-joint pliers to bend these stubs inward or outward around the outermost wire to form an opening with no sharp projections. This is important because if you cut the stubs off, even with flush-cutting wire cut-

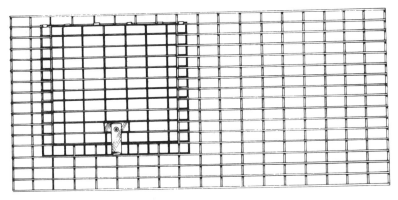

*Door overlaps opening by 1" on sides and bottom and swings inward.*

ters, the opening would be sharp. The result would be scratched hands, arms, and rabbits. There are other ways to do it: You could cut them off flush and file them smooth or cover them with flanges of sheet metal. I find leaving stubs and bending them over to be less time-consuming and more satisfactory.

**Step 7. Attaching the Door.** For the door, cut 12″ of 1″ × 2″ wire from the 36″-wide roll. Measure down 14″ and cut with all the cuts flush. While you are at it, measure down another 14″, cut another door for the next hutch, and set it aside. You will have a piece 6″ × 12″. Cut the stubs off that and off the end of the roll. Set the small piece aside.

Before you attach the door, attach the latch. Position it up 2″ from the bottom of the center of the door. It fits over the strands 2″ apart. If you lay it on a bench or the floor you can flatten it easily over the wire with your hammer. If you wait until it's on the door to do this, you will have a difficult time. With the latch on, fit the door *inside* the door opening and clip it with J-clips to the top of the opening.

Use a J-clip on each end and 3 across the middle. Do not give them such a tight squeeze and the door will swing freely. Your door will overlap the sides and bottom by an inch, and the latch should work easily. Give it a drop of oil if it doesn't.

Now swing the door to the top of the cage (the ceiling) and, 5″ over from the left edge of the top, squeeze the door hanger onto the top of the cage with your slip-joint pliers. When the cage is upright, the hanger will hold the door up while you reach into the cage. Give the door a push and it will swing loose and shut. The beauty of this door is that it is always inside the cage, not out in the aisle to snag your sleeve. Even if you forget to latch it, your rabbits cannot escape; it will stay shut no matter how hard they push it.

**Step 8. Add a "Baby Saver."** The hutch now looks finished, and in fact it is usable, but it needs some finishing touches. To prevent baby rabbits from falling out of the hutch if they fall out of the nest box, a "baby saver" is needed.

Take the 2"-wide strip you set aside when you split the 36" wire for front and back sections. "Stagger" it over one end at the bottom and fasten it with J-clips every 6". This will close the openings at the bottom to 1/2" × 1", up 2-1/2" and will prevent the babies from falling through. Cut a 2" strip for the front and another for the back and fasten them on. When you split another front and back section for the next hutch, clip that one on the other end. When you get to the end of the roll, you will find that you will be able to build 10 hutches in this manner and have enough material for the baby saver feature. In the meantime, save the pieces from the door openings and doors for later use as hay racks. We'll discuss them later.

## Building More Than 10 Hutches

If you plan to build more hutches — even if you don't plan to build them all now — the following is a better plan. It will build up to 36 hutches.

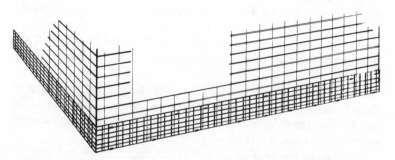

*Adding the "baby saver" feature — 1/2" × 1" wire fastened around the bottom of the sides of the hutch.*

Buy a roll of 1″ × 2″ "side" wire, 100′ long, either 18′ or 15′ wide (depending upon whether you raise medium or small rabbits) for every 9 hutches you plan to build (4 rolls for 36 hutches). Also buy a 100′ roll of 30″-wide 1″ × 2″ "top" wire, a 100′ roll of 36″-wide 1/2″ × 1″ floor wire, and a 50′ roll of 12″-wide 1″ × 2″ "door" wire.

**Building Plan.** Measure and cut off 11′ of the side wire. Form three 90° corners at 30″ and 36″ intervals around your 2 × 4 and fasten the remaining corner with J-clips. Cut off 3′ of the 30″ top wire and fasten it on. Cut off 42″ of the 36″ floor wire. With your 2 × 4 and hammer, bend up 3″ at 90° on all 4 sides, cutting out the corners, to form a box. You have now built the baby saver feature right into the floor. Because it is all one piece, the hutch will be considerably stronger this way. Fasten the floor as described earlier, but also fasten the baby saver sides to the front, back, and ends of the hutch. Make the doors as described earlier.

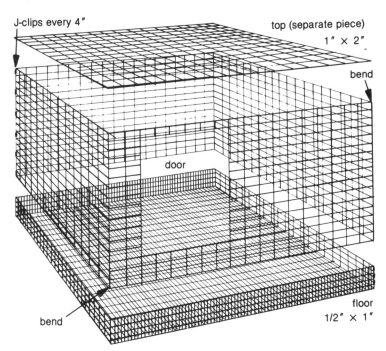

*Cutting plan for an all-wire hutch when you build 10 or more.*

If you decide to build multiple-hutch units with shared partitions, use the 1/2" × 1" wire to divide them. Otherwise, the rabbits may fight through the partitions. Order your wire supply accordingly.

# Self-Feeders

To equip this hutch completely, you will want to attach a self-feeding hopper. Farm stores and rabbitry supply houses have them in various sizes. The trough should be positioned about 4" above the floor for small and medium breeds. The hopper portion remains outside the hutch for easy filling by the rabbit keeper. The trough is inside. It takes up no floor space, and it is high enough and narrow enough to keep the rabbits from fouling it.

Successful rabbit raisers have abandoned crockery feeders (which cost as much or more anyway, and often break) for these self-feeders, which attach with 2 spring wire hooks on each side and, of course, detach easily for periodic cleaning. Tin cans have also gone the way of the crock — they are out. Self-feeders have a lip that prevents the rabbits from scratching pellets out, leaving them hungry and the keeper poor from buying more feed to be wasted. Hopper self-feeders are worth the money and in the long run would save you cash over tin cans even if they were made of sterling silver.

*A typical self-feeder — hopper variety. This feeder clips to the outside of the hutch and can be filled from outside.*

# Hay Racks

You can make 2 kinds of hay racks easily with the small scraps of 1 × 2 wire that remain from door openings. Both are simple to make, so you have no need to purchase manufactured hay racks.

For the first type, simply take a 6–8" square scrap and bend 2" of it to an acute angle of about 30°. With a couple of J-clips, fasten the 2" side to the front of the hutch wherever it is convenient, even on the door. Fill the rack with a handful of hay and the rabbits will pull it through.

Another type of hay rack is especially useful for feeding alfalfa hay, which has leaves that tend to fall from the stems when pulled by the rabbits. Build it about 6–8" over the trough of the feeder on

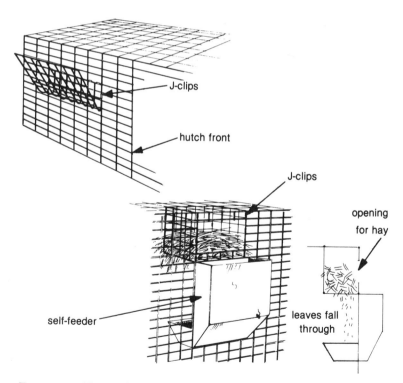

*Two types of hay racks: simple all-purpose rack, and alfalfa hay rack.*

the inside of the hutch. You simply form a box of 1″ × 2″ wire and J-clips that projects 2″ into the hutch up to the roof and as wide as the feeder trough. Cut a slot in the front of the cage the width of the rack minus a couple of inches, and about 3″ high. Any hay leaves that the rabbits don't eat on the first attempt will fall into the feed trough for a second chance.

# Watering Devices

Rabbits require a constant supply of fresh water to grow and maintain health. Rabbit raisers say water is the best food, because without it a rabbit will not eat properly .

Supplying that water can be the most laborious task in rabbit raising — or you can make it so simple that it practically takes care of itself. Let's review the ways you can supply water to rabbits.

**A Tin Can or Other Open Dish.** Very poor, as the rabbit will tip it over and go thirsty until you refill it. Difficult to clean, it invites disease. And there is a lot of work involved in refilling if you have many rabbits.

**A Crock.** Better, because it can't be tipped. If you use crocks, get the kind with a smaller inside diameter at the bottom than at the top. During freezing weather, the water will expand into ice and slide up. Otherwise it will expand out and break the crock. The problem with crocks is that they are, like the tin can, open and susceptible to fouling by the rabbits. They are also labor-intensive, with lots of refilling and washing.

**A Tube Bottle Waterer.** The plastic bottle with the drinker tube is a big improvement over the can or crock. The enclosed water supply stays clean. No space is taken from the cage floor. They require less washing than crocks. They will not work in freezing weather, but ice doesn't break them.

**A Plastic Bottle With Drinker Valve.** You make this one yourself because, amazingly, nobody manufactures one. All it takes is a large heavy plastic bottle, such as one for bleach or soda pop, and a drinker valve designed for automatic watering systems. The bottles

bleach jug
with drinker valve

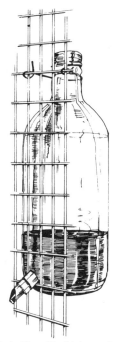

soda bottle with drinker valve

tin can

crock

tube bottle waterer

*Manual rabbit-watering devices.*

are free (unless they are returnable for a deposit); the valves cost a dollar or so and are available at farm supply or rabbitry supply houses.

With a knife or drill bit, make a hole in the bottle near the bottom, as shown in the drawing. Coat the threads of the valve with epoxy cement and screw or push it in. Use wire as shown to hold the bottle onto the cage. A quart or half gallon bottle works well and supplies plenty of water, although you can use more than one per hutch if necessary. The plastic bottle has all the advantages and disadvantages of the tube bottle except that it is cheaper and more durable, can supply more water, and, if large enough, can cut filling time considerably.

**A Semiautomatic Watering System.** This is a fine way to water a small rabbitry. You need a tank, which need be no more than a 5-gallon jerry can or pail, leading via flexible or rigid plastic pipe to drinker valves at each hutch. You can buy almost everything locally, with the possible exception of the valves, which were developed for use by poultry.

A very simple setup utilizes flexible black plastic pipe that can be fitted out with various adapters, couplings, elbows, and tees. The

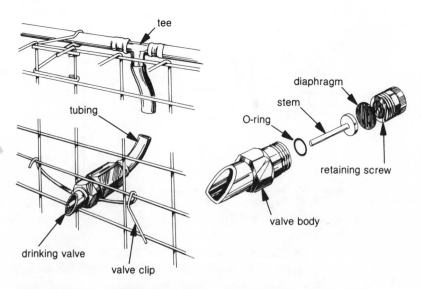

*Semiautomatic watering system with detail of drinking valve.*

pipe runs along the outside of the cage, where it cannot be gnawed, about a foot or so above the hutch floor. Using a simple hand-tapping tool, you make the holes for the valves, which screw in and protrude into the wire cage. The rabbits quickly learn to drink from these valves: their licking dislodges a brass tip, letting the water spill into the rabbits' mouths. Another type of valve has a spring-activated stem that opens when bitten and closes when released. The latter are better because they rarely leak.

At least one manufacturer, Borak Ltd., produces semiautomatic watering systems as kits, but you can put your own together from various components in farm supply, hardware, or plumbing supply stores.

The beauty of the semiautomatic watering sytem is that you merely fill the holding tank (ideally with a garden hose) and the rabbits drink a constant supply of fresh water that remains clean and takes up no floor space in the hutch. In addition, you can keep this water flowing in the coldest weather by inserting electric heating cables inside the pipes. You can buy these cables in various lengths, depending on how many cages you have, and they can be operated manually or with a simple air-temperature-activated thermostat. Most of them deliver 2-1/2 watts per foot but some produce more heat. Because they are inside the pipe, they need to produce very little heat to keep the water flowing. In severe climates you should use two such cables inside the pipe, activating one with a thermostat and the other manually when the temperature really drops. Without heating cables, you can keep the pipes from freezing by draining them at night. The Borak system is engineered so that you can conveniently disconnect the whole system, if you have a small rabbitry, and take it indoors during freezing nights. All rabbit raisers should consider the semiautomatic system.

**An Automatic Watering System.** This ultimate system is simply the semiautomatic system with a piped water supply from a well or city water system. It requires the equipment of a semiautomatic system as well as a means of reducing the pressure before water enters the plastic supply pipe to the valves. One way to reduce pressure is a float valve in the tank (like a toilet tank). As the rabbits drink, the float valve allows more water to enter the tank, keeping it constantly full. Pressure-reducing valves are also available. You can keep your float valve and tank system function-

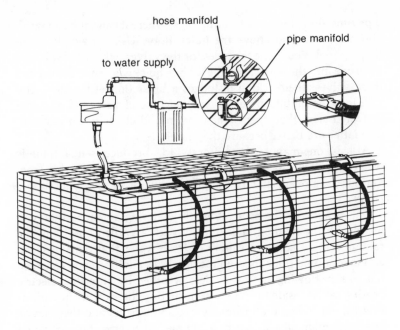

*Automatic watering system with detail of pressure-reducing and filtering equipment.*

ing in cold weather by coiling a length of heating cable inside the tank.

The rabbitry supply houses listed at the back of this bulletin will provide all the materials you require for your watering system if you send them a simple dimensional diagram of your hutch layout.

While some rabbit raisers may be put off by the plumbing and electrical aspects of such a system, you can put one together quite easily. Once installed it takes all the hard work — the hauling and pouring of water — out of caring for rabbits.

To sum up the watering situation, while it is common to begin with crocks or bottles, a piped system can actually cost you less per hutch if you have a large number of hutches and under any circumstances costs very little more. It is important, however, to have all your hutches in place before installing the system, particularly if you will need heating cables, which cannot be shortened or lengthened.

# Nest Boxes

Place a nest box in an all-wire hutch on the 27th day after you mate the doe. The litter will be born in it on the 31st day, ordinarily. You must use one because of the open nature of the all-wire hutch. There are basically 3 kinds of nest boxes to choose from.

## Types of Nest Boxes

You can make your own nest box from wood. Be sure to cover all the edges with metal; if you do not, they will be gnawed to nothing. For the medium breeds, make the dimensions 10" × 18" of floor space and 8" high. Cut the front down to about 4" for easy use by very young rabbits. Be sure to drill 6 or 7 drainage holes in the floor. Don't put a cover on it as it will become damp. After each use a wooden nest box must be washed, disinfected, and left to dry in the sun.

You can buy ready-made galvanized metal nest boxes. They have removable hardboard floors, but because the tops are partially enclosed, these boxes often become damp. Dampness endangers rabbit health. It's also difficult to see into these boxes for daily inspection of the litter.

You can buy or build all-wire nest boxes with removable corrugated cardboard liners. This is the type of nest box I have used exclusively for many years and strongly recommend. The box is made with 1/2" × 1" floor wire and J-clips, and it has metal flanges covering the top edges to protect the rabbits from injury as they hop in and out.

In very cold weather, use it with the corrugated cardboard liner, perhaps with an extra layer of corrugated cardboard or foam plastic to insulate the floor. Use a new liner, free of any possible germs, for each litter and destroy the liner later. You can cut these corrugated liners from boxes, usually obtainable free.

In warm weather, cut cardboard for only the floor and use a shallower bedding of shavings and straw. Leaving the wire mesh sides open gives plenty of ventilation, which is very important to guard against dampness and extreme heat, two rabbit killers. The open mesh provides you a fine view of the litter's daily progress.

## Building the All-Wire Nest Box

For the medium-size and smaller breeds, cut out a floor section of 1/2" × 1" 16-gauge galvanized wire that is 10" × 18". Fourteen-gauge wire will do, but it is not necessary. Cut out a back section 8" × 10" and side sections 8" × 18". For the front, cut an 8" × 10" piece and cut a V-shape notch as shown. Use J-clips 4" apart to fasten the box together. Having built an all-wire hutch or two, you will find this a simple task.

Cover the sharp edges along the top with flanges of galvanized metal that will also serve to clamp the corrugated liner in place. With a pair of tin snips or other metal cutters, cut the flanges from 28-gauge galvanized sheet metal (available from building and heating supply dealers). For the sides, make them 3" wide × 17" long. For the back, 3" × 9". The front requires 2 flanges to fit the V-shape notch. Make them 3" × 5" each. Cut the ends at the angle shown.

A simple way to form the flanges into the required shape is to bend the sheet metal pieces around a 1/2" wooden dowel, available in hardware and building supply stores. Wear work gloves to avoid cutting your fingers. Nail a piece of 1/4" wood lath that is 2" or 3"

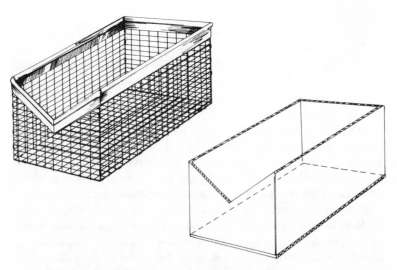

*The all-wire nest box with corrugated cardboard liner.*

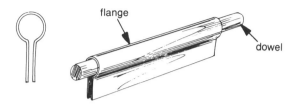

*Dowel makes a keyhole shape in metal flange.*

wide and a couple of feet long to a workbench. If you haven't got a workbench, nail the lath to a heavy plank and place that on the floor or a table.

Lay a piece of the flange-to-be on the lath, with 2" off and 1" on, lengthwise. Place the dowel on the sheet metal at the edge of the lath and hammer a slight curve into it by striking the dowel along its length. Then, with your thumbs on the metal, use your fingers to pull the sheet metal around the dowel. Once you have pulled it all around, lay on it a short piece of board (a piece of 1" × 2" furring strip or other scrap will do) and strike it with your hammer. Turn the piece of metal over and continue striking the board until you have the desired shape. Doing this takes less time than telling about it. Repeat the operation for all the flanges.

Cut your cardboard liner an inch smaller than the inside dimensions of the box and score the folds with a knife. Bend it to the box shape and fit it into the box. Then cut out a "second floor" of 1/2" × 1" wire that is 9" × 17" and place it inside the box to prevent the rabbits from scratching through the cardboard.

In warm weather, poke 6 or 7 drainage holes through the box floor. In very warm weather, cut cardboard for the floor only, poking drainage holes through it. You will use no cardboard sides in hot weather, so twist pieces of flexible wire around the flanges and through the side wire to hold the flanges on. You will not need to do this when using the full cardboard liner.

In very cold weather place additional cardboard or foam plastic on the wire floor, add a layer of shavings and insert the second wire floor. A good source of this foam plastic is the supermarket. Fresh meat and vegetables often come in plastic foam trays. The shavings will absorb moisture below the second floor. Of course, use several inches of shavings and plenty of straw above the second floor, filling the nest box to the top. The doe will burrow into it to make a nest for the litter and will line it with fur from her own body.

This versatile nest box will last a lifetime. It needs no special cleaning because a new liner is used each time. It works well in all kinds of weather. It is light and compatible with the all-wire hutch and can be stored right on top of the hutch. Furthermore, if you assess your cage requirements carefully when ordering floor wire, it is entirely possible you can build your nest boxes from floor wire remaining on a roll if you have bought it 100' or even 50' at a time.

A parting word about nest boxes: There are 2 kinds you should never use. First, some suppliers are offering a plastic cat bed as a nest box. Rabbits will eat plastic. Second, do not try to use a cardboard box alone. The doe will tip it over — if she doesn't eat it first. Either way, the litter will surely be lost.

# The Carrying Cage

Sooner or later you will want to transport your rabbits to market, to sell, or to show. The all-wire carrier or carrying cage is the best way to do so. It is lightweight and will keep the rabbits cool and well ventilated during transportation.

 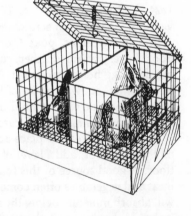

*Rabbit carrying cages.*

You can build yours from small pieces of 1" × 2" and 1/2" × 1" wire. After building cages and nest boxes you will find it a cinch. Dimensions of the floor should give the rabbit enough room to stretch out. Measure a stretched-out rabbit and start from there. For short hauls, the height need be no more than 8". For longer trips, which would require feeding and watering along the way, you will need more height, and, in fact, a junior-size hutch.

Put your carrier together much like the nest box except use 1" × 2" wire for the sides and top and 1/2" × 1" for the floor. (You can use 1/2" × 1" wire all around if you have it.) Raise the floor 1" from the bottom so there is clearance underneath to provide for drainage and droppings enroute. Cut the top an inch or two larger than the floor and bend the corners to fit over the sides. Hinge it from the back with J-clips, and use a dog leash snap fastener or similar clasp to clip it closed.

A metal pan underneath the carrier will catch droppings and urine. Or you can find plastic trays or dishpans in housewares departments of local stores and build your carriers to fit them. Put an inch or so of shavings in the tray and attach it to the carrier with short hooks of wire or small springs.

# More Useful Tools and Equipment

In addition to the tools you used for hutch construction, you will find several other tools valuable in rabbit raising.

• A wire brush with a long handle, as used by house painters, will loosen droppings that occasionally stick on hutch floors. After prolonged use the brush will wear thin on the end. Saw off the worn bristles to give the brush new life.

• A propane torch with a flame-spreading head will burn off hair that sticks to hutches. This hair can catch droppings, so periodic use of the torch will help keep hutches clean. (Unless you want to cook them in the hutch, remove the rabbits first.)

• A manure fork comes in very handy for removing manure to garden or compost pile or to bag up for sale at the local garden center.

• A Garden Way Cart surpasses a wheelbarrow in many ways, but it really shines in rabbit raising. It will make short work of manure removal and also can be used for carrying heavy bags of feed and bales of hay, straw, and shavings. Buy the large Model 26 unless your doorways will admit only the medium-size Model 20. I use both in my rabbitry.

• A hanging scale, to which you can attach a pail or an all-wire nest box, is excellent for weighing young rabbits. Keeping track of the growing rabbits' weight progress is important if you want to produce only good gainers in your rabbitry.

• A photographer's aluminum floodlight reflector with a low-wattage bulb can be placed on top of an all-wire hutch to shine warmth into an open nest box when a litter is expected during extremely cold weather.

• A pelt stretcher (which doesn't really stretch the pelt but does dry it for tanning) is made of heavy spring steel wire. Buy them from rabbitry supply houses or make your own from 6-gauge galvanized wire.

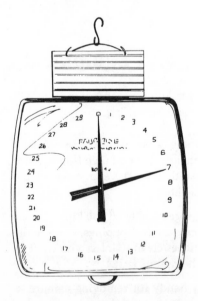

A hanging scale for weighing rabbits.

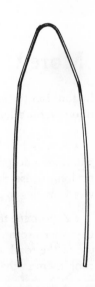

A pelt stretcher for drying rabbit pelts.

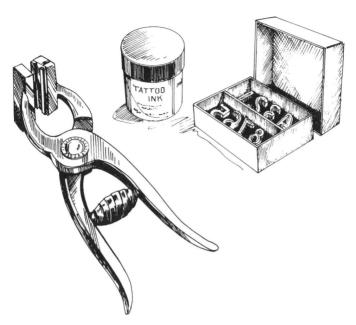

*A tattoo set for permanently marking rabbits' ears.*

• A tattoo set, consisting of tattoo pliers or tongs, ink, an ink brush, and a set of numbers and letters will permit you to keep track of breeding rabbits by marking their ears permanently. Your private ear number goes in the rabbit's left ear. This number is required if you show your rabbit. Registered rabbits are marked in the right ear by the registrar.

• A galvanized garbage can, the regular 26-gallon size, will hold 100 pounds of pellets and keep them clean and dry and safe from vermin.

• A feed scoop from your farm supply store beats using a tin can.

Finally, please avoid one piece of equipment often seen in rabbitries — the salt spool hanger with its spool. Salt spools tend to drip and rust holes through hutch floors. If you feed rabbit pellets there is no need to add salt. If you must add salt, shake it from a shaker into the feeder.

# Additional
# Sources of Information

More information on rabbit raising is available from books, organizations, and suppliers. For current addresses of any source listed below, send your request to Bob Bennett, One Governor's Lane, Shelburne, Vermont 05482. Include a self-addressed stamped envelope.

## Books

The following books are ones I have found useful. For a list of current sources and prices, write me at the above address. Include a self-addressed stamped envelope and ask for the book list.

*Raising Rabbits the Modern Way*, by Bob Bennett. A complete guide to raising rabbits on a small or semicommercial scale from Garden Way Publishing.

*Bob Bennett's Guide to Winning Rabbit Shows.* All about getting into showing rabbits. 66 pages with photographs. 1977.

*The T.F.H. Book of Pet Rabbits*, by Bob Bennett. Completely illustrated with full-color photographs and of special interest to serious rabbit raisers and pet owners alike because of the pictures, taken by the world's best photographers of rabbits and beautifully printed. 77 pages. 1982.

*Rabbit Production*, by Cheeke and Patton of the Oregon State University Research Center, is an update of *Domestic Rabbit Production* by Templeton. Chapters were contributed by Lukefahr and McNitt, also researchers. Textbookish and technical but has much information worth the effort of reading. 1982.

*The Private Life of the Rabbit*, by R. M. Lockley. The author is an English naturalist and writes only about wild rabbits but provides insights not found elsewhere. 200 pages, hardcover. 1973.

*How to Start a Commercial Rabbitry*, by Paul Mannell. Apparently written to promote sales of his company's equipment, which is advertised therein, but has handy insights nevertheless. 100 pages, soft cover.

*Practical Inbreeding*, by W. Watmough. What inbreeding does for all livestock. Good for a beginner. 67 pages.

*The Book of the Tan Rabbits*, by A. S. Howden. The author bred the ancestors of some of my Tans. He provides specific information on Tans as well as general good breeding and management tips. 50 pages.

*The Book of the Dutch Rabbit*, by James Read. Does for Dutch what Howden does for Tans. 32 pages.

*Modern Angora Wool Farming*, by Carl Nagel. Information on angoras by a former raiser. 90 pages.

*Green Foods for Rabbits and Cavies*. What greens you can feed and what greens to avoid. 76 pages.

*Rabbit Raising*, by Bob Bennett. This is the official Boy Scout Merit Badge book. Aimed at youngsters, it has tips for every new raiser. Inexpensive and handy to give to new breeders.

*Official Guide Book of the American Rabbit Breeders Association, Inc.* 200 pages of essential information written by amateur volunteers. Free with membership in the ARBA. To join and receive your book, as well as *Domestic Rabbits* magazine, which was originated by Bob Bennett, send $10 for one year's membership to Bob Bennett, One Governor's Lane, Shelburne, Vermont 05482.

## Newsletters and Guidebooks

For each recognized breed there is a newsletter and a guidebook from the specialty club devoted to the particular breed. Membership is required, at about $5 per year, with additional benefits to be gained. Addresses for these clubs can be obtained by sending a stamped, self-addressed envelope to the author, address above.

## Breeding Stock

If you are unable to locate breeding stock of the desired breed, you may write the author for his recommendations of U.S. and Canadian breeders. Enclose a stamped, self-addressed envelope and state the breed or breeds in which you are interested.

## Suppliers

The following firms supply a complete line of rabbitry equipment. I have made purchases from them and know them to be reputable.

Safeguard Products, 114–116 Earland Drive, Dept. B, New Holland, Pennsylvania 17557. Best hutch designs. Complete equipment line. Most progressive company in the business.

Michi-Crown, RR 2, Alger, Michigan 48610. Send $1.00 for catalog, which is refunded with first order. Lots of handy items.

Glick Manufacturing Co., 420B East 9th St., Gilroy California 95020. Send $2.00 for "Commercial Rabbit Raisers Guide" with complete catalog.

New England Rabbitry Supply, RFD 3B, North Middleboro, Massachusetts 02346. Send $1.00 for their catalog. It is worth the money for the disease control information. Owner Joe Laura is a rabbit judge.

Mountain View Rabbitry Supply Co., Naples, Maine 04055.

Morton Jones Company, PO Box 123, 925B Third Street, Ramona, California 92065. Send for their free catalog.

Klubertanz's, 1165B Highway 73, Edgerton, Wisconsin 53534.

Borak Equipment Inc., 4325B Nordum Road, Everson, Washington, 98247.

Jewell Enterprises, Star Route, Box 264B, Detroit Lakes, Minnesota 56501.